The Brain Behind the Eyes

How Your Brain Helps You See and Understand the World

By Dr. Rafaela Thé

Published by: Dr. Rafaela Thé

First Edition, March 2026

Library and Archives Canada Cataloguing in Publication data available upon request.

Welcome to The Brain Behind the Eyes!

In the first book of the Vision Learning Series, **The Eye Book**, you learned about how amazing your eyes are! In **The Brain Behind the Eyes**, you'll discover how truly amazing your brain and eyes are when they work together to help you see the world, store memories, solve problems, and so much more!

Get ready to take a peek inside the most powerful organ in your body, your **brain**!

Whether you're a deep thinker, a big dreamer, or a curious question-asker, there's something in here for you!

Table of Contents

Meet Your Brain!

Your **brain** is the boss of your whole body! It helps you think, feel, move, and even **see**. Everything you do, like reading this book, remembering your friend's name, or kicking a ball, happens because of your brain.

Your brain is made up of billions of special cells called **neurons.** These tiny cells send messages to each other at lightning speed, like a super-fast game of telephone. Your brain uses these messages to help you make choices, solve problems, and understand the world around you.

Even though your brain weighs only around **3 pounds** (about the size of a small watermelon), it is the most powerful part of your body. It works all day and all night, even when you sleep!

Let’s Meet the 3 Main Parts of the Brain:

1. The **Cerebrum** is the largest part of your brain. It helps you think, learn, and remember things. The cerebrum has two halves, called the **left and right hemispheres**. The left side helps with things like math and language, while the right side is great at imagination and creativity. But there’s more! Each half of the cerebrum has four main parts, called **lobes**, and each one has a special job:

   - **Frontal lobe:** is at the front of your brain, right behind your forehead. It helps you **make decisions**, **solve problems**, and **control your movements**. It is also where your **personality** lives!
   - **Parietal lobe:** is located near the top of your head. It helps you make **sense of touch**, **temperature**, and **pain**.
   - **Temporal lobe:** is located on the sides of your brain, near your ears. It helps you understand **sound** and **language.** It is also important for **memory**.
   - **Occipital lobe:** is located at the back of your brain. Its main job is to **help you see**! It is also where you’ll find the **visual cortex,** which is the part of your brain that turns light and images from your eyes into pictures you can understand.

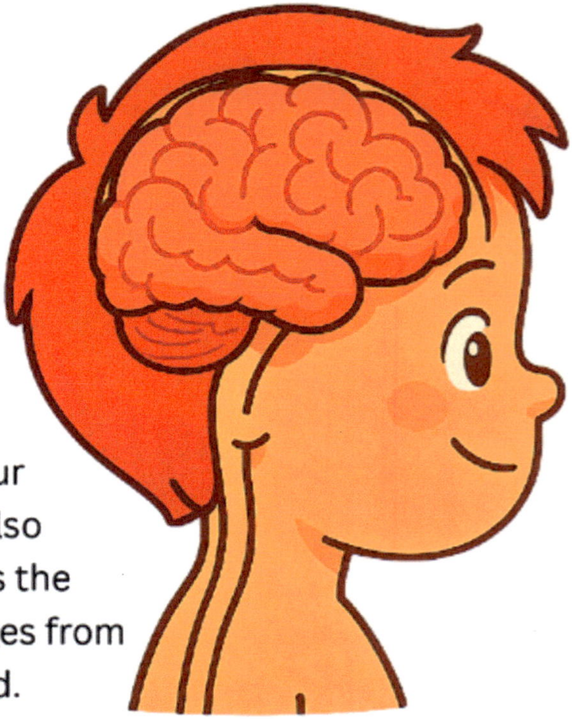

Each lobe works together as a team, to help you do everything from reading a book, to kicking a soccer ball, or dreaming up a story.

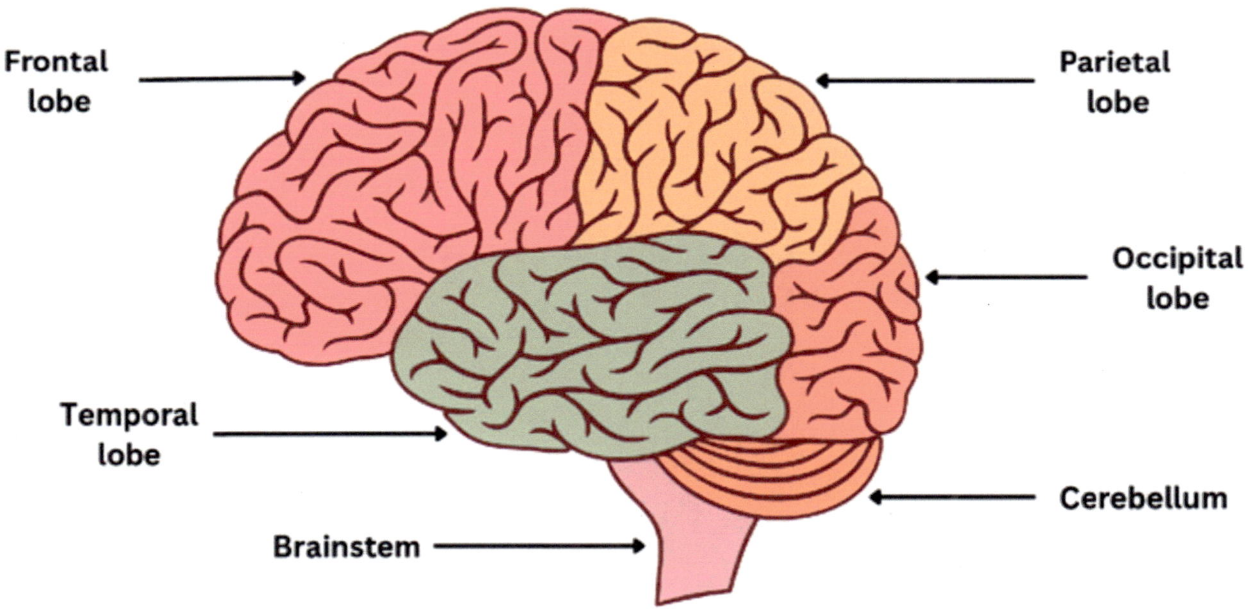

2. The **Cerebellum** sits under the cerebrum. It helps you **balance, move smoothly,** and **stay coordinated**. So, if you are dancing, riding a bike, or catching a ball, you can thank your cerebellum!

3. The **Brainstem** links your brain to your **spinal cord**. It controls things you don't even have to think about, like your **heartbeat, breathing**, and **digestion**. It keeps everything running without you ever noticing.

Fun Fact

Your brain tells the rest of your body when something hurts, BUT the **brain** itself **can't feel pain**! Isn't that amazing?

Did You Know?

Your brain keeps changing, learning, and growing throughout your life. This is called **neuroplasticity** (neuro-plas-Ti-city), and it means you can always get better at thinking, remembering, and creating, no matter how old you are!

Brain Booster: Try This!

Close your eyes and try tickling yourself. Does it work? **No!**

That's because you **can't tickle yourself!** Your brain can predict the sensation and turns down the response, so you don't feel tickled.

Your brain is so smart!

Eyes and Brain: A Perfect Team

Your **eyes** are amazing, but did you know they **don't actually see**? That's right! Your eyes collect information, but it is your **brain** that turns it into **vision**. Your eyes and brain are a team working together every second of the day!

How Your Brain Sees the World:

When you look at something, light enters your **eye** and passes through the **cornea**, **pupil**, and **lens**. These parts focus the light onto the **retina**, a special layer at the back of your eye.

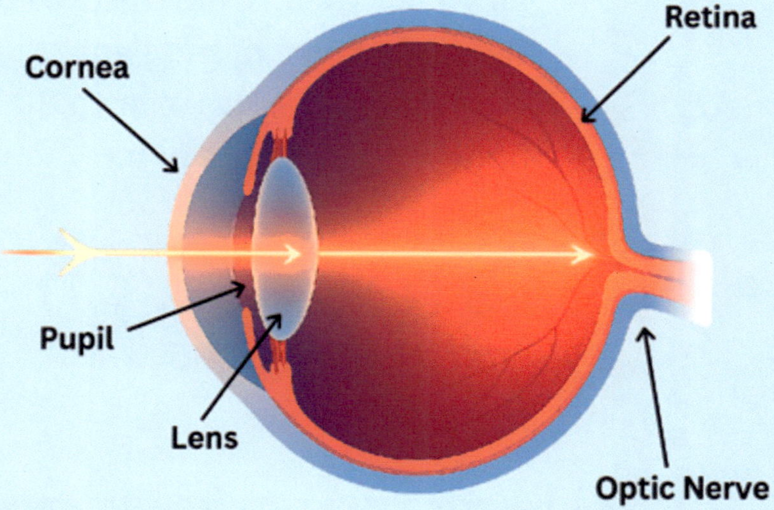

The **retina** is covered in tiny cells called **photoreceptors** that turn light into signals. These signals carry the visual information to the back of your brain, to an area called the **visual cortex** (in the occipital lobe).

The **visual cortex** then puts everything together and figures out what you are seeing, like a tree, a friend's face, or a flying soccer ball.

And the most amazing part, this entire process happens in **less than one second!**

Fun Fact

As signals travel from the eyes to the brain, some of the signals from each eye **cross over** to the opposite side of the brain.

This helps both sides of your brain share information and work better together.

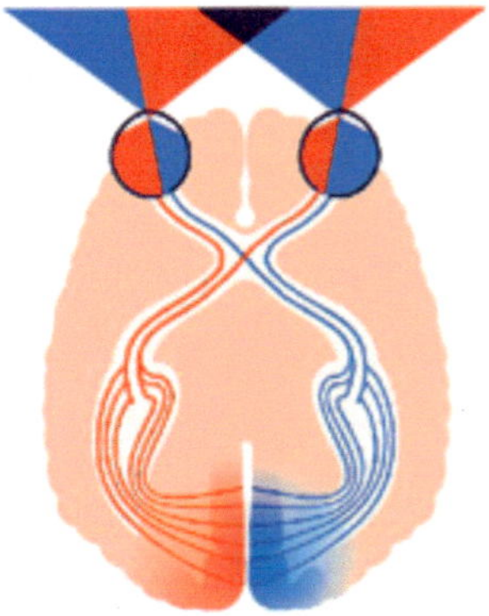

Did You Know?

Your brain uses more energy to help you see than it does on anything else you do while you're awake.

About 40% of your brain is involved in vision!

Brain Booster: Try This!

Stand on one foot and close your eyes. Is it more difficult to balance? **Yes!**

That's because your brain uses vision to help with balance. Without vision, your brain is relying only on your inner ear and muscles to balance.

Growing Brains, Growing Eyes

When you were born, your brain and eyes were already working together, but they had a lot of growing to do. Just like you've gotten taller and stronger, your eyes and brain have gone through some big changes, too.

Let's take a look at how vision develops from the very beginning of life.

Age	What's Happening?
0 – 1 month old	- Vision is very fuzzy when babies are born. - Babies can see only about 20-30 cm (~8-12 in) in front of them. - They see mostly black, white, and grey. - Babies have no 3D vision (depth perception) at birth.
2 months old	- Babies start being able to follow moving objects with their eyes. - They begin to develop hand-eye coordination.
3 – 4 months old	- Babies start to develop colour vision. - Red is the first colour babies start to see. - 3D vision starts to develop.
5 - 6 months old	- Babies can now mostly see across a room. - They can recognize familiar people from far away. - Vision continues to get better. - Their colour and 3D vision is improving.

Crawling is important for the eyes and brain:

Did you know that crawling is really important for the eyes and vision to develop properly? Here's why:

Skill	
Eye-Hand Coordination	Crawling encourages babies to look ahead, reach, and coordinate movements between what they see and how they move.
Depth Perception (3D Vision)	Moving across the floor gives babies experience judging distances, which is really important for developing depth perception.
Binocular Vision	Crawling helps the eyes learn to work together to follow objects and maintain focus while they are moving.
Tracking Skills	As babies crawl toward toys or people, they practice smoothly following objects with their eyes and changing focus between near and far things.

Research shows that babies who **crawl longer** tend to score higher on **visual perception** and **eye-hand coordination** tasks later in childhood.

Depth Perception: Learning to See in 3D

Depth perception means being able to tell how far away something is. When babies are born, they only see things in 2D, which is how we see when watching TV. As babies start to move around, depth perception (3D vision) starts to develop.

When babies begin crawling, reaching, and grabbing for things, their brain gets better at using both eyes together to understand space and distance. That's when they start to avoid bumping into furniture or reaching too far for a toy.

Did You Know?

Most of your brain's growth happens before age 6, but areas that control vision, memory, and decision-making continue developing for many years.

Brain Booster: Try This!

Look at a family member or friend and try to notice something new about their face, like the shape of their eyebrows or the the shape of their smile.

Your brain gets better at recognizing faces as you grow. This helps you spot someone you know in a crowd.

Seeing is Believing, or is it?

Have you ever looked at something and thought it was something else, like a pile of clothes that looked like a person? That's your brain playing a trick on you. Sometimes, what you think you see isn't really there.

Your brain doesn't just snap a picture like a camera. It has to figure out what you're seeing by using shapes, shadows, colours, and past experiences. This is called **visual perception**, and is how your brain understands what you are seeing.

But here's the thing, your brain will sometimes take shortcuts. It fills in missing details and makes guesses to help you recognize things quickly. Usually, that's super helpful, but sometimes it causes silly mix-ups, like thinking a rope is a snake or reading a word that is not actually there.

Visual Perception

Without looking back, can you think of anything weird about the mascot on the last page?

Ok, now go back and look carefully. Do you see it now?

Your brain quickly makes sense of the picture with what it knows is "normal" even for a cartoon. That's why you probably didn't see it had three hands the first time around.

Optical Illusions:

Optical illusions are special pictures or designs that confuse your brain by using tricky shapes, patterns, or colours. You might see things move, change size, or even appear when they're not really there.

Although optical illusions are fun, they also teach us something important. Your brain does not just show you what is in front of you; it helps you understand what you are seeing.

Brain Booster: Try This!

Which Red Circle is Bigger?

Look at the two red circles. One is surrounded by tiny black circles, and the other is surrounded by big circles

Do the red circles look like they're different sizes?

Most people think the red circle on the left looks bigger than the one on the right. Guess what? They're actually the exact **same size**!

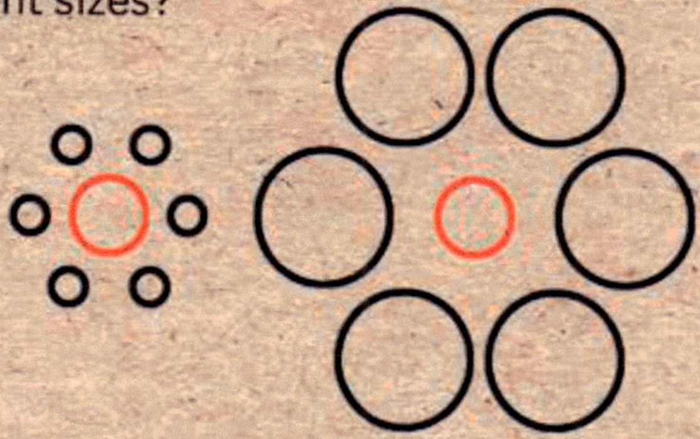

What's Going On?

This is an optical illusion called the Ebbinghaus illusion. Your brain gets tricked by what is around the red circle.

That's because your brain compares things to what is surrounding it. So, when the red circle is next to the small ones, it looks larger by comparison. When the red circle is next to the large ones, it seems smaller.

Your brain is just trying to help; it uses what's around to figure out size, but sometimes, that causes a mix-up!

Brain Booster: Try This!

Which Line is Longer?

Look at this image with the two red lines. Which line is longer? The two red lines are the actually the **same length**, but one looks longer than the other.

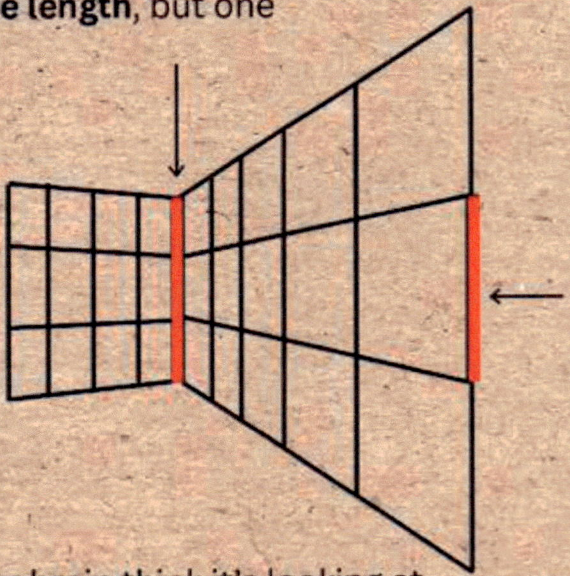

What's Going On?

This picture shows a special kind of optical illusion called the **Ponzo illusion**. Your brain is very good at using clues in a picture to figure out how far away something is.

In this illusion, the background makes your brain think it's looking at something that is 3D (like the corner of a room). When your brain thinks an object is farther away, it adjusts the size in your mind to make sense of it. That's why it looks longer, even though both lines are actually the same size.

Brain Booster: Try This!

Curved or Straight?

Take a look at the two red lines. Do they look curved or bent to you?

They look like they are bending away from the middle slightly, but they are actually perfectly **straight**!

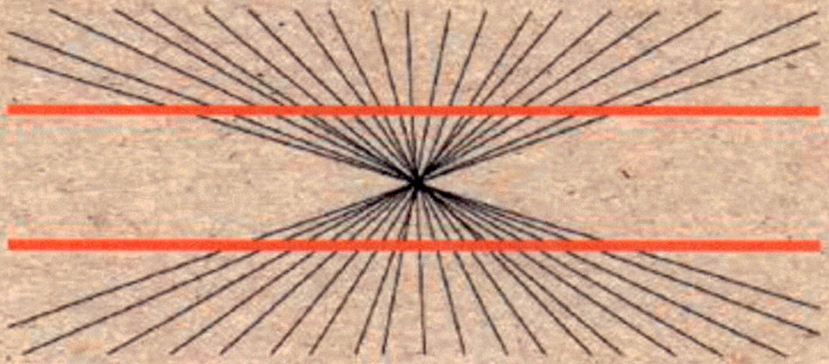

What's Going On?

This is called the **Hering illusion**. Your brain is being tricked by all the black lines in the background. Those black lines are pointing outward like sun rays, and they confuse your brain into thinking the red lines are bending away from the middle.!

Your brain tries to make sense of all the lines together, and that makes the straight red lines look curved, even though they are not!

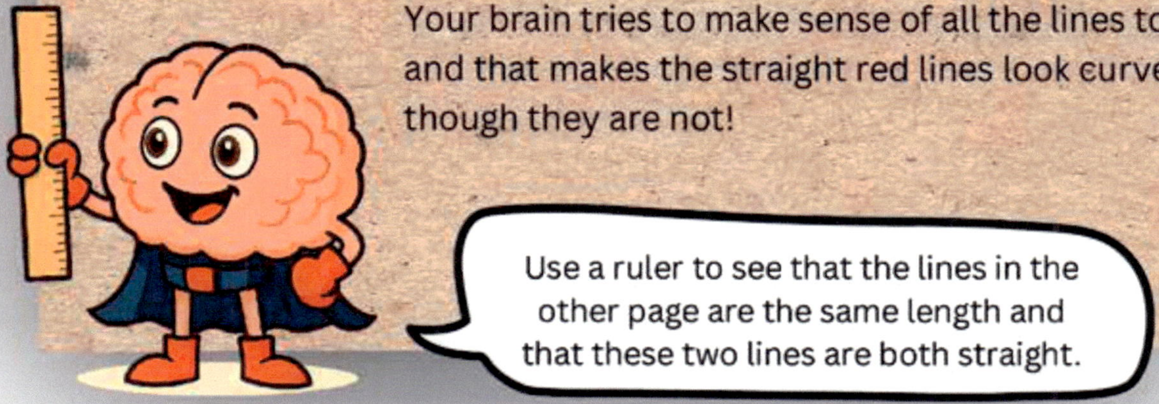

Brain Booster: Try This!

Is the Blue Side in the Front or the Back?

You see this line drawing of a cube? Look at it for a few seconds and it may flip. Sometimes the blue side looks to be in the front of the cube and other times it looks like it is in the back of the cube.

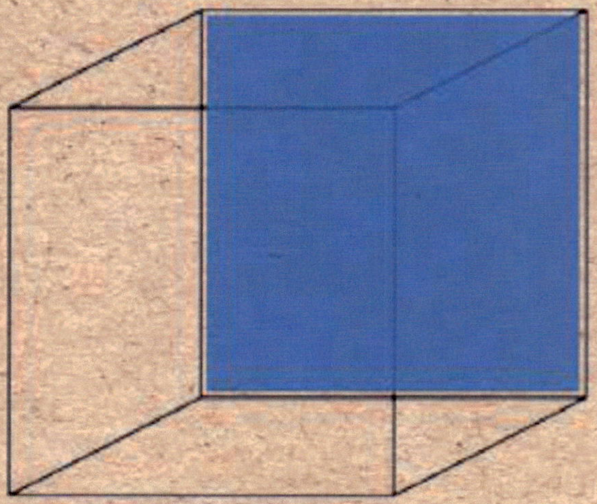

What's Going On?

Your brain is trying to make sense of a 2D drawing of a cube and interpret it as a 3D object. Because this drawing has no shadowing or other clues to help your brain decide if the blue side is in front or the back, both options make sense. Your brain can't choose just one answer, so it switches between them. This is called a **"bi-stable" image**.

Brain Booster: Try This!

Same Blue... or Not?

Look at the two blue circles on the cube. Does one look lighter than the other? It might seem like the left circle is darker and the right one is brighter. Here's the cool part: they are actually the **same colour**!

What's Going On?

This is called a **shadow and light illusion.** The cube makes your brain think one side is in the shadow and the other side is in the light.

Your brain is super smart and tries to "fix" the coloured circles to make sense of the light and shadow surrounding them. Based on your brain's past experiences, it knows that light and shadow affect colours differently. It thinks the circle in the shadow must be lighter than it looks, and the one in the light must be darker. So it changes how you see them, even though the colour hasn't changed at all!

Movement and Focus

Every time you read, play sports, or walk through a busy hallway, your **brain** and **eyes** have to **work together** to keep things **clear** and **steady**. That teamwork helps you stay balanced, track movement, and stay focused on what's important.

Let's break it down:

- There are **6 muscles** around each eye that control **eye movements**. These muscles move your eyes up, down, and from side to side, allowing both eyes to work together. Your brain sends messages to those muscles to guide them where to look!

- When you follow a moving object, like a bouncing ball or a flying bird, your brain controls **tracking**. This helps you smoothly follow motion so the image does not look jumpy or blurry.

- When you switch your focus from something far away (like the board at school) to something close up (like your notebook), your brain helps your eyes **refocus** using a skill called **accommodation.** The lens inside your eye changes shape to help focus at a new distance and your brain fine-tunes the picture, so you can still see clearly!

Good eye movements and focusing skills are also important for learning in school. If your eyes have trouble moving smoothly across a page or jumping from one line to the next, reading can feel tiring or confusing. You might lose your place, skip words, or have trouble understanding what you just read. That's why eye doctors check how well your eyes move, not just how clearly you can see.

Sometimes people will need **vision therapy** or eye exercises to teach their eyes and brain how to work together properly.

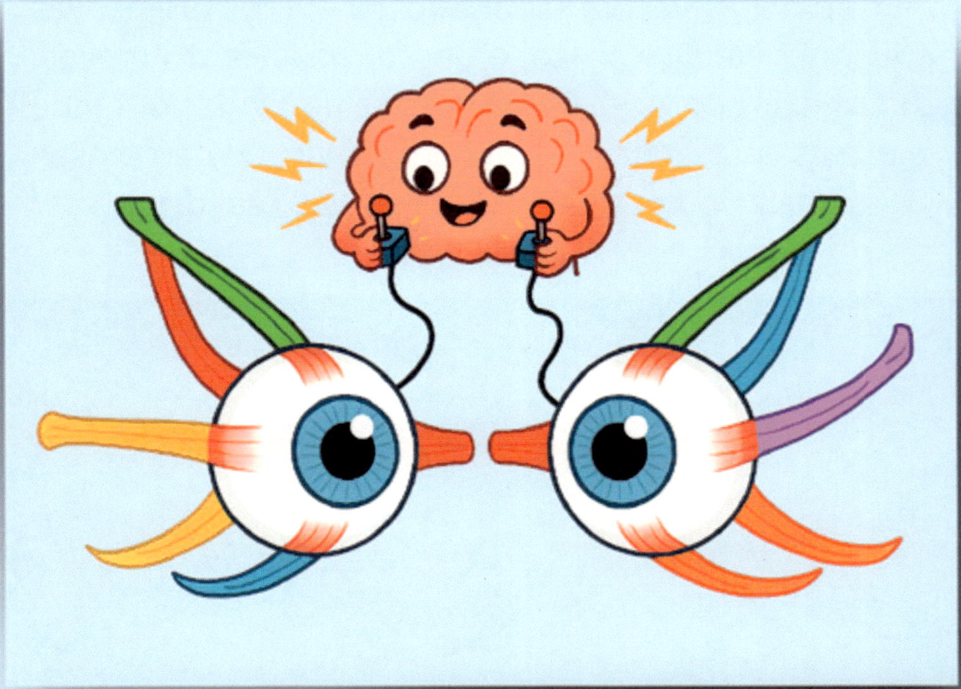

The better your brain and eyes work together, the easier it is to read, move, play, and explore the world around you. And just like any winning team, the more your brain and eyes practice working together, the better they get!

Fun Fact

Your eyes also help you **balance**! If you've ever felt dizzy when spinning, you've experienced how the brain, eyes, and inner ear all work together for balance. Your eyes send signals about what you are seeing, while your inner ear tells your brain how you are moving. When those signals do not match, it makes you feel **dizzy**.

Did You Know?

Your eyes make tiny jumps and movements about 3-4 times every second. That adds up to over 200,000 movements a day. They will keep moving even when you are asleep and dreaming!

Brain Booster: Try This!

Hold your thumb out at arm's length and stare at it. Now move your thumb slowly from side to side. Can your eyes follow it smoothly? **Yes!**

That's your **tracking** skills at work, which is controlled by your brain!

Memory and the Mind's Eye

Have you ever pictured your favourite toy, even when it's not in front of you? That's your brain using your **mind's eye**! It is how you imagine or remember things you've seen, like a friend's face, a dream, or picturing an object spinning in your mind.

Your brain is always storing memories, like how your room looks or what your best friend's smile looks like. These memories help you recognize people, places, and even signs or logos you see all the time. This is called **visual memory**, and it is one of your brain's coolest superpowers.

Your brain stores different kinds of memories:

- **Short-term memory** holds small bits of information just long enough to use them, like repeating a phone number in your head until you write it down. Short-term memory usually lasts about 15-60 seconds.

- **Working memory** helps you hold and use information, like remembering steps in a math problem or following directions. It lasts a little longer and involves **active thinking**.

Working memory is closely related to **short-term memory** but involves **active thinking**. That means your brain isn't just holding information; it's doing something with it, like following instructions while you build something.

- **Long-term memory** keeps important information for hours, days, or even a lifetime, like your address, how to ride a bike, or remembering your 6th birthday party. Long-term memories take time to build and usually need practice, repetition, or strong emotions.

- **Visual memory** is especially important for reading, spelling, finding your way around, or remembering what things look like. If you have ever drawn a picture from memory or remembered how to get to the park without a map, that is your visual memory in action!

Everyone's memory works a bit differently. Some people have a strong **visual memory;** they remember best by seeing pictures, words, or colours. Others have great **auditory memory** and learn well by listening to sounds, music, or voices. Some people rely on **kinesthetic memory**, which means they remember better by moving, building, or acting things out. Most people use a **mix of all three** types of memory, and that's what makes learning so interesting and unique for everyone!

You can improve your memory by practicing. Games like matching cards, drawing a picture from memory, and even making up silly rhymes or songs can help make it stronger. Getting good sleep and eating healthy foods also give your brain what it needs to remember better.

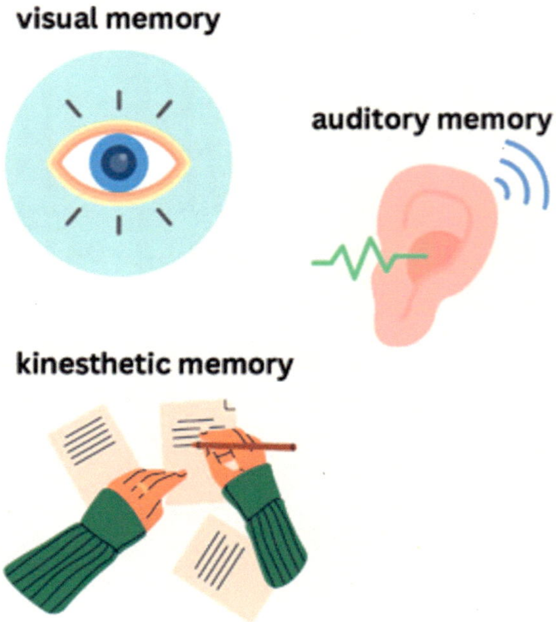

Did You Know?

Your visual memory even works when you are dreaming. That's why dreams often have pictures!

Brain Booster: Try This!

Look at a picture for 30 seconds. Then close the book and try drawing it from memory.

How much can you remember?

That is your **visual memory** in action!

Protecting Your Brain and Eyes

Your **brain** and **eyes** do incredible things, but they need your help to stay healthy and strong! Just like wearing a jacket in the cold or sunscreen in the sun, there are smart ways to protect your eyes and brain every day.

Head Safety:

Your skull acts like a helmet for your brain, but during activities like biking or sports, you still need extra protection. Wearing a **helmet** helps prevent serious injuries if you fall or bump your head.

Wearing a helmet can **lower** the risk of serious head injury by up to **85%** during a fall or crash!

Nutrition:

Healthy foods fuel your brain and eyes. Colourful fruits and veggies (like carrots, spinach, and blueberries), whole grains, fish, eggs, and nuts are full of vitamins and nutrients that help your eyes see clearly and your brain think better.

Outdoor Play:

Spending time outside every day is really important for your whole body, but also for your eyes. Did you know that spending time outside is helpful to **reduce** the risk of becoming **nearsighted**. Doctors recommend spending about 2 hours outside per day.

Movement:

Exercise gets your heart pumping, which sends fresh oxygen to your brain and eyes, helping you stay alert, focused, and strong.

Sun Safety:

Sunlight has **UV rays**, which are invisible light rays that can hurt your eyes. **Sunglasses** with UV protection and wide-brimmed hats act like shields to keep your eyes safe.

Screen Time Tips:

Too much screen time can make eyes feel tired or dry. This is called **digital eye strain**. To help, follow the 20-20-20 rule: every 20 minutes, look at something 20 feet away for 20 seconds.

Blinking more often and keeping screens about an arm's length away also helps!

Rest:

Sleep gives your brain and eyes a chance to recharge. While you sleep, your brain stores memories and solves problems. Your eyes get a break, too!

Kids aged 6 to 12 need about **9 to 12 hours** of sleep every night (depending on age).

Did You Know?

Your brain is about **75% water**. So remember to stay hydrated to help you think more clearly.

Brain Booster: Try This!

Make a **BRAIN JAR!**

Each time you learn something new, write it on a piece of paper and add it to a jar. Watch your brainpower grow as you learn!

Quiz: How Well Do You Know Your Brain?

1. **Your brain works even while you are asleep.**
   ☒ True
   ☒ False

2. **What part of the brain helps you balance and move smoothly?**
   A) Cerebrum
   B) Brainstem
   C) Cerebellum

3. **Which part of your eye sends messages to your brain?**
   A) Retina
   B) Optic nerve
   C) Cornea

4. **Your brain stops growing and changing once you become a teenager.**
   ☒ True
   ☒ False

5. **Fill in the blank: Your brain is made up of special cells called _________.**

6. **What is the name for when your brain guesses what you are seeing and sometimes gets tricked?**

7. **Which of the following helps protect your brain?**
   A) Sunglasses
   B) Helmet
   C) Sandals

8. **What should you do every 20 minutes during screen time to protect your eyes?**

9. **What is the name of your brain's "photo album" that helps you remember what things look like?**

10. **Your brain can recognize a familiar face in less than one second.**
    ☒ True
    ☒ False

Glossary

Accommodation - The way your eye changes focus from something far away to something close-up and from near to far.

Active Thinking - Using your brain on purpose to solve problems, make decisions, or figure things out. It's like when you really focus and think hard to answer a tricky question.

Auditory Memory - Remembering things by hearing them, like sounds, music, or spoken words.

Brain - The control center of your body! It helps you think, learn, feel, remember, move, and even see. Your brain works like a supercomputer that is always turned on, even when you're sleeping.

Brainstem - The part of the brain that controls automatic things like breathing, heartbeat, and digestion.

Cerebellum - The part of the brain that helps with balance, coordination, and smooth movements.

Cerebrum - The largest part of your brain. It helps you think, learn, speak, and use your five senses.

Cornea - The clear, dome-shaped window at the front of your eye. It helps focus light so you can see clearly.

Depth Perception (3D vision) - Your brain's way of figuring out how far away things are. It helps you catch a ball, pour juice into a cup, or walk down stairs without tripping!

Digital Eye Strain - Tired, sore, or blurry eyes after looking at screens for a long time.

Eye - The part of your body that lets you see the world around you. Your eyes take in light and send messages to your brain so you can understand what you're looking at.

Eye-Hand Coordination - When your eyes and hands work together to help you do things like catching a ball, writing your name, or building with blocks. Your brain uses what your eyes see to guide your hands where to move.

Kinesthetic Memory - Remembering things by moving your body or using your hands. Also called motor or tactile memory.

Lens - A clear part inside your eye that changes shape to help you see things up close or far away.

Mind's Eye - The part of your brain that helps you "see" pictures in your head, even when your eyes are closed.

Neurons - Tiny cells in your brain and body that send messages to help you think, move, and feel.

Neuroplasticity - Your brain's ability to change, grow, and make new connections as you learn.

Optic Nerve - The nerve that carries messages from your eyes to your brain so you can see.

Optical Illusion - A trick picture or pattern that confuses your brain and makes you see something that is not really there.

Photoreceptors - Special cells in your retina that turn light into signals for your brain.

Pupil - The black circle in the middle of your eye. It gets bigger or smaller to let in just the right amount of light.

Retina - The back part of your eye that collects light and starts the seeing process.

See - To "see" means to use your eyes and brain together to understand what's around you. Your eyes collect the light, and your brain turns it into pictures so you know what you are looking at.

Short-term Memory - A type of memory that holds information for a short time, like remembering a phone number for a few seconds.

Spinal Cord - A long bundle of nerves inside your backbone that sends messages between your brain and the rest of your body.

Tracking - When your eyes follow something that moves, like a flying ball or words across a page. Your brain and eye muscles work together to keep your eyes on the target!

UV Rays (Ultraviolet Rays) - A type of invisible light that comes from the sun. Too much UV can hurt your eyes and skin, which is why wearing sunglasses and using sunscreen is so important!

Vision - How your eyes and brain work as a team to help you see the world. It includes things like seeing shapes, colours, movement, and even how far away something is.

Vision Therapy - Special eye exercises that help your eyes and brain work better together. It's like a workout for your vision, helping with things like focusing, eye tracking, and seeing clearly while reading or playing!

Visual Cortex - The part of your brain that helps you understand what your eyes are seeing.

Visual Memory - The ability to remember what things look like, such as faces, places, or pictures. It's like the "photo album" of the brain.

Working Memory - The part of your brain that helps you hold and use information at the same time, like solving a math problem in your head.

Visual Perception - How your brain understands what your senses, especially your eyes, are telling you.

Quiz Answers and Review

1. **Your brain works even while you are asleep.**
   Answer: True! Your brain is always working, storing memories, solving problems, and helping your body rest.

2. **What part of the brain helps you balance and move smoothly?**
   Answer: C) Cerebellum

3. **Which part of your eye sends messages to your brain?**
   Answer: B) Optic nerve

4. **Your brain stops growing and changing once you become a teenager.**
   Answer: False! Your brain keeps growing and changing throughout your life! This is called neuroplasticity.

5. **Fill in the blank: Your brain is made up of special cells called ________.**
   Answer: Neurons

6. **What's the name for when your brain guesses what you are seeing and sometimes gets tricked?**
   Answer: Visual perception or optical illusion.

7. **Which of the following helps protect your brain?**
   Answer: B) Helmet

8. **What should you do every 20 minutes during screen time to protect your eyes?**
   Answer: Follow the 20-20-20 rule (look at something 20 feet away for 20 seconds)

9. **What is the name of your brain's "photo album" that helps you remember what things look like?**
   Answer: Visual memory or the mind's eye

10. **Your brain can recognize a familiar face in less than one second.**
    Answer: True!

Keep Learning with the Vision Learning Series

Your brain and eyes work together in amazing ways! Want to keep exploring?

The Eye Book
Learn all about how your eyes work and why they are so important.

The Light Book (Coming Soon)
Find out what light is and how it helps your eyes and brain work together.

The Eye Activity Book
Try fun activities, puzzles, and games to explore vision in a hands-on way!

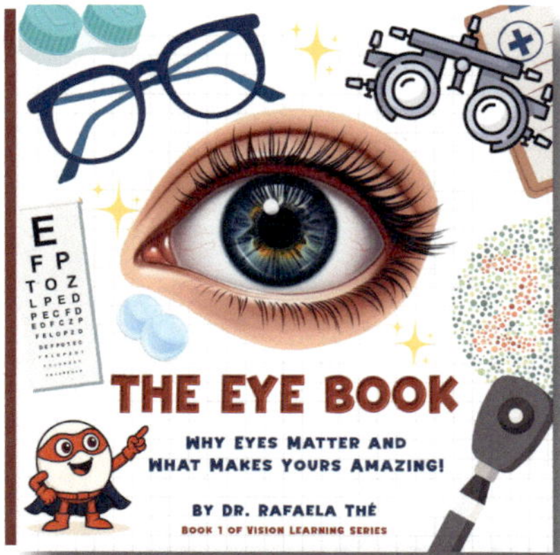

Made in the USA
Monee, IL
05 November 2026

67381892R00024